HOW TO DESIGN AND BUILD A PHASE CONVERTER

TABLE OF CONTENTS

M&D CO.PHASE CONVERTER

=================================

CHAPTER 1

ENCLOSED ARE THE DIAGRAMS AND EXPLANATIONS TO
RUN THREE PHASE MOTORS ON SINGLE PHASE

THESE SHOULD BE SIMPLE ENOUGH FOR MOST TO
FOLLOW BUT IF YOU HAVE ANY PROBLEMS JUST EMAIL
US AT (chigeor@bellsouth.net).

BELOW IS A PARTS LIST AND WHERE YOU CAN GET
THE PARTS NEEDED TO BUILD A PHASE CONVERTER.

YOU SHOULD BE ABLE TO FIND ALL THE PARTS
NEEDED AT ANY AIR CONDITIONING SUPPLY HOUSE.
ELECTRICAL SUPPLY HOUSES STOCK THE "K2" RELAY.

THE PARTS NEEDED ALSO CAN BE FOUND AT WWW.
GRAINGER.COM AND THEIR PART NUMBERS ARE WHAT
WE HAVE LISTED BELOW.

YOU CAN TAKE THESE NUMBERS AND ANY SUPPLY
HOUSE CAN CROSS REFERENCE THEM FOR YOU TO
THEIR PART NUBERS.

K2 IS PART NUMBER 5X847, A STANDARD RELAY WITH
ONE SET OF "NORMALLY CLOSED CONTACTS" RATED
AT LEAST 30 AMPS AND HAS A 120 VOLT COIL.

K1 RELAY IS PART NUMBER 2A560 AND WE RECOMMEND
GETTING THIS FROM WWW.GRAINGER.COM . WE HAVE
HAD GOOD LUCK WITH THIS ONE.

CAPACITORS ARE DIFFERENT SIZES FOR DIFFERENT
HORSEPOWER OF MOTORS.

BUT A 4X662 PART NUMBER IS A GOOD ONE TO START
WITH.

A GOOD RULE OF THUMB IS TO USE ONE 40-50MFD
CAPACITOR FOR EVERY ONE HP OF THE MOTOR.

SO, A THREE HP MOTOR WOULD NEED A TOTAL OF
ABOUT 120-150 MFD FOR THE STARTING CAPACITORS.

CAPACITORS HOOKED UP TO EACH OTHER IN PARALLEL
ADD TO EACH OTHER AND HOOKED IN SERIES THEY
SUBTRACT FROM THE TOTAL MFD.

IF YOU DECIDE TO TRY TO BALANCE THE OUTPUT
VOLTAGE OF YOUR MOTOR , YOU WILL USE WHATS
CALLED RUNNING CAPACITORS WHICH WILL BE
DISCUSSED LATER. THE SAME RULE APPLIES HERE,
ABOUT A 40-50 MFD FOR EACH HP OF THE MOTOR.

CHAPTER 2

THREE LEAD MOTORS:
======================
LABEL YOUR TWO LINES THAT FURNISH 220 VOLTS AS
LINE "A" & "B".

YOU ALSO MUST HAVE A NEUTRAL OR COMMON LINE
WITH THESE.
IT DOES NOT MATTER WHICH IS WHICH , "A" OR
"B".
NEXT ON THE MOTOR, ALSO LABEL THESE THREE
LEADS "A", "B", "C".

IT DOES NOT MATTER WHICH IS LABELED AS SUCH. NOW CONNECT ONE WIRE FROM YOUR 220 VOLT LINE "A" TO THE MOTOR "A".

THEN CONNECT THE "B" 220 VOLT LINE TO MOTOR "B"
THIS LEAVES ONLY "C" OF THE MOTOR LEFT TO CONNECT.
ATTACH A WIRE FROM MOTOR LEAD "C" TO ONE SIDE OF THE

CAPACITOR C2.   IT MAKES NO DIFFERENCE WHICH TERMINAL OF THE CAPACITOR YOU USE.

YOU WILL THEN CONNECT THE OTHER TERMINAL OF THE CAPACITOR "C2" TO A TERMINAL SCREW OF THE "K2" RELAY.

THIS TERMINAL SCREW MUST GO TO THE NORMALLY CLOSED SET OF CONTACTS.

THESE CONTACTS ARE CLOSED WHEN NO POWER IS APPLIED.

FOLLOW THE POINTS TO THE TERMINAL SCREW TO MAKE SURE YOU HAVE THE RIGHT TERMINAL.

NOW THE OTHER SIDE OF THESE CONTACTS, YOU WILL RUN A WIRE FROM THAT TERMINAL SCREW AND ALSO CONNECT IT TO THE LINE "A" OF THE INCOMING 220 VOLTS WE USED EARLIER.

THE NEXT STEP IS TO TAKE THE TIME DELAY RELAY "K1" AND ATTACH THE OUTPUT SIDE OF THE RELAY TO THE NEUTRAL OR COMMON WIRE.

ADJUST THIS RELAY COUNTER CLOCKWISE UNTIL IT STOPS AND THEN TURN CLOCKWISE TO AN ADJUSTMENT OF ONLY 1/2 A SECOND.   NO MORE!!!

THERE ARE ONLY TWO TERMINALS ON THIS RELAY, VERY SIMPLE.

ATTACH A WIRE FROM THE INCOMING TERMINAL OF "K1" OVER TO A TERMINAL SCREW ON THE COIL OF RELAY "K2".   THE OTHER

TERMINAL SCREW OF THE COIL ON "K2" WILL CONNECT TO LINE "B" OF THE INCOMING 220 VOLTS.

YOU CAN ADJUST THE TIMING OF "K1" A LITTLE AT A TIME UNTIL THE MOTOR STARTS PROPERLY.   BUT NOT OVER ABOUT ONE SECOND MAXIMUM.

THE WAY IT WORKS IS,   WHEN YOU APPLY 220 VOLTS TO THE MOTOR, THE TIME DELAY RELAY IN ABOUT 1/2 A SECOND,   OPEN THE CONCACTS ON RELAY "K2" AND REMOVE THE CAPACITOR OR CAPACITORS FROM THE CIRCUIT.

THIS ALLOWS THE MOTOR TO RUN ON ITS OWN.

THE STARTING CAPACITORS MUST COME OUT OF THE CIRCUIT IN THIS LENGTH OF TIME OR THEY WILL BLOW.

CHAPTER 3

NINE LEAD MOTORS:
=====================

FIRST YOU HAVE TO LOOK AT THE NAMEPLATE OF THE MOTOR AND CONNECT THE NINE LEADS FOR 220VOLTS.

NOT 440VOLTS. STANDARD CONNECTION IS LEADS 4-5-6 ARE TIED TOGETHER.

THEN 1-7 TOGETHER AND 2-8 TOGETHER, AND 3-9 ARE TOGETHER.
LABEL THE 220 VOLT INCOMING LINES AS "A" & "B".

IT MAKES NO DIFFERENCE WHICH IS LABELED AS SUCH.

NOW GO BACK TO THE MOTOR AND LABEL THE TWO LEADS TIED TOGETHER 1-7 AS "A".

AND THE 2-7 AS "B", AND 3-9 AS "C".

NOW CONNECT WIRE "A" FROM THE 220 LINE TO "A" OF THE MOTOR.

THEN DO THE SAME WITH "B" OF THE 220 LINE TO "B" OF THE MOTOR.

THIS LEAVES US WITH 4-5-6 AND THE "C" LEFT TO CONNECT.

ATTACH THE MOTOR LEAD "C" TO ONE SIDE OF CAPACITOR "C2" AS IN DIAGRAM.

EITHER SIDE OF THE CAPACITOR WILL WORK, IT MAKES NO DIFFERENCE WHICH.

NOW THE OTHER SIDE OF "C2" WILL GO TO THE TERMINAL SCREW OF THE "K2" RELAY. THIS SCREW MUST BE ON THE NORMALLY CLOSED SET OF CONCACTS, SO LOOK AT THE RELAY CLOSE AND BE SURE.

THIS SET OF CONTACTS CLOSED WHEN NO POWER IS APPLIED ANYWHERE.

THEN TAKE THE OTHER SIDE OF THE NORMALLY CLOSED SET OF CONTACTS AND RUN A WIRE TO LINE "A" OF THE 220 VOLTS COMING IN.

FOLLOW THE DIAGRAM AND ITS PRETTY SIMPLE TO SEE.

NOW LOOK AGAIN AT "K2" RELAY AND YOU WILL SEE THE COIL AND IT ALSO HAS TWO TERMINAL SCREWS.

RUN A WIRE FROM EITHER ONE OF THESE TO THE 4-5-6 LEADS ON THE MOTOR AND TAPE THESE UP.

THE OTHER SIDE OF THE COIL YOU WILL CONNECT A WIRE FROM THAT TERMINAL TO LINE "C" OF THE MOTOR.

FOLLOW THE DIAGRAM AND YOU WILL HAVE NO
PROBLEM.

THE HORSEPOWER OF THE MOTOR WILL DETERMINE THE
SIZE OF THE CAPACITOR OR CAPACITORS "C2" THAT
YOU NEED.

A GOOD RULE OF THUMB IS TO USE 40-50 MFD FOR
EVERY ONE HORSEPOWER OF THE MOTOR.   A THREE
HP MOTOR WILL TAKE ABOUT 120-150 MFD TO run.

START IT.   A FIVE HP WILL NEED ABOUT 200-250
MFD TO START IT.   A SIMPLE FORMULA BUT ITS A
GOOD STARTING POINT

THE SIZE VARIES WITH DIFFERENT MOTOR
MANUFACTURERS BECAUSE OF THE AMOUNT OF IRON
THEY USED INSIDE THAT THE WINDING USES TO
MAGNETIZE TO DEVELOP HORSEPOWER.

CHAPTER 4

HOW TO RUN ONE CONVERTER FOR THE ENTIRE SHOP:

THIS IS THE BEST WAY TO RUN THREE PHASE MOTORS
ON SINGLE PHASE VOLTAGE.

WHAT YOU NEED TO DO FIRST IS TO DETERMINE THE
LARGEST HORSEPOWER MOTOR YOU WANT TO RUN OFF
OF THIS CONVERTER.

THEN THE ROTARY CONVERTER YOU WILL BUILD NEEDS TO BE LARGER THEN THIS MOTOR.

IF YOUR ROTARY CONVERTER IS A 10 HP MOTOR YOU CAN RUN A 7 1/2 HP MOTOR OFF OF IT AND A TOTAL OF TWO MORE FIVES FOR EXAMPLE.

JUST DO NOT TRY TO START MORE THEN ONE AT A TIME.

START ONE AND THEN YOU CAN START THE NEXT ONE.

DO NOT OVERLOAD THE CONVERTER MOTOR.

IF THE MOTOR YOU WANT TO USE FOR A ROTARY CONVERTER HAS NINE LEADS, THEN USE THE DIAGRAMS ENCLOSED FOR THAT MOTOR AND BUILD THE CONVERTER AS SHOWN.

WHEN THIS IS DONE AND YOU HAVE THE CONVERTER MOTOR RUNNING, YOU USE THE LEADS OF "A", "B", "C" FOR THE THREE PHASE VOLTAGE TO SUPPLY TO OTHER MOTORS.

ITS A VERY SIMPLE SETUP.
AFTER THE CONVERTER IS RUNNING, YOU HAVE ESSENTIALLY BUILT A THREE PHASE GENERATOR THAT CAN BE USED TO START YOUR OTHER THREE PHASE MOTORS AS LONG AS ITS LARGER THEN THEY ARE.

YOU SHOULD RUN THESE LEADS "A", "B", "C" TO
A 3 PHASE BREAKER PANEL AND THEN HAVE
INDIVIDUAL BREAKERS FOR EACH MOTOR YOU WANT TO
RUN.

IF YOU WANT TO RUN ANYTHING FROM A 120 VOLT
LINE OF THIS CONVERTER, SUCH AS A LIGHT BULB
OR WHATEVER. 'DO NOT TIE IT TO THE "C" LEAD.

THAT VOLTAGE VARIES AND WILL BURN UP
WHATEVER YOU HOOK TO IT.

YOU CAN ALSO BALANCE THE VOLTAGE OF "C" USING
A VOLTMETER AND ADDING WHAT IS CALLED RUNNING
CAPACITORS CONNECTED FROM LINE "C" TO LINE "A"
AND LINE "B".

START WITH THE SAME FORMULA AS BEFORE. ABOUT
40-50MFD FOR EVERY HP OF THE MOTOR AND WATCH
THE VOLTAGE ON LINE "C" READING TO GROUND.

WHEN ITS PRETTY CLOSE TO THE SAME VOLTAGE AS
"A" AND "B" YOU HAVE THE RIGHT SIZE RUNNING
CAPACITORS.

ITS A TRAIL AND ERROR UNTIL YOU GET IT CLOSE
BECAUSE OF THE DIFFERENT MANUFACTURES OF
ELECTRIC MOTORS. BUT THIS IS CLOSE.

CHAPTER 5

CAPACITORS:

THERE ARE TWO TYPES OF CAPACITORS USED FOR
CONVERTERS.

ONE IS CALLED A STARTING AND THE OTHER IS
CALLED A RUNNING CAPACITOR.

STARTING CAPACITORS ARE USUALLY LARGER IN
WHATS CALLED MFD AND HAVE A LOT OF KICK TO
THEM TO JUMP START THE MOTOR.

THE RUNNING CAPACITORS IF USED, HELP THE
CURRENT FLOW OF THE THIRD THREE PHASE LEG THAT
IS BEING GENERATED BY THE CONVERTER MOTOR.

THE STARTING CAPACITOR MFD SIZE YOU USE IS
DETERMINED BY THE HORSEPOWER OF THE MOTOR YOU
ARE GOING TO OPERATE.

REMEMBER TO START WITH ABOUT 40-50 MFD FOR
EVERY ONE HORSE POWER OF THE MOTOR.

YOU CAN PARALLEL TWO CAPACITORS AND THEY WILL
ADD UP TO THE TOTAL MFD OF BOTH.   IF YOU
SERIES THEM,   THE TOTAL WILL BE SMALLER THEN
THE SMALLEST ONE.
A ROTARY PHASE CONVERTER CAN BE STARTED AND
RUN WITH ONLY RUNNING CAPACITORS BUT IT TAKES
A LOT OF THEM TO ACCOMPLISH THIS.   BUT THEN
IT ELIMINATES SWITCHES AND TIMERS AND RELAYS
BUT IT IS DONE THIS WAY AT TIMES.

YOU WILL HAVE TO PLAY WITH THE SIZE OF THE
CAPACITORS BUT USING THE SIMPLE FORMULA ABOVE
WILL GET YOU IN THE BALL PARK AND YOU MAY THEN
NEED TO GO UP OR DOWN WITH THE SIZE OF THE
CAPACITORS.

NOT ALWAYS IS MORE BETTER. TOO MUCH MFD WILL
NOT START THE MOTOR JUST AS TOO LITTLE WILL
NOT EITHER.

CHAPTER 6

STARTING THE MOTOR:

WHEN YOU APPLY 220 VOLTS TO A 3 PHASE MOTOR
WITH A CONVERTER , THIS IS WHAT HAPPENS.

YOUR 220 VOLTS IS APPLIED TO "A" AND "B" OF
THE 3 PHASE MOTOR.

THIS MOTOR HAS THREE WINDINGS IN IT AND YOU
HAVE APPLIED VOLTAGE TO ONLY TWO OF THEM. IT
WILL NOT START.

THE CONVERTER THEN TAKES A VOLTAGE FROM LINE
"A" AND RUNS IT THROUGH A CLOSED SET OF
CONCACTS AND THROUGH THE STARTING CAPACITORS.

THESE CAPACITORS ARE TIED TO THE OTHER LINE
"C" OF THE MOTOR AND GIVE IT A VOLTAGE KICK TO
START THE MOTOR TURNING.

WHEN THE MOTOR REACHES FULL SPEED, WHICH
SHOULD HAPPEN IN LESS THEN A SECOND IN MOST
CASES.

THE CONTACTS THEN OPEN ON "K2" AND REMOVE THE
STARTING CAPACITORS FROM THE CIRCUIT.

THE MOTOR IS NOW RUNNING ON SINGLE PHASE ONLY.

ON THREE LEAD CONVERTERS THAT USE THE "K1"
TIMER RELAY, THIS IS CONTROLLED BY THE
SETTING OF THE TIMER.

DO NOT SET IT OVER ONE SECOND ON MOST MOTORS
ARE THE STARTING CAPACITORS WILL BLOW FROM
STAYING IN THE CIRCUIT TOO LONG.

THEY CAN'T TAKE MUCH LONGER THEN THAT.

A RUNNING CAPACITOR STAYS IN THE CIRCUIT IF
THEY ARE USED AND ALSO CAN HELP TO START THE
MOTOR.   THEY ALSO HELP BALANCE THE OUTPUT
VOLTAGE OF LINE "C" WHEN THE RIGHT SIZES ARE
USED.

USE THE SAME FORMULA FOR THESE RUNNING
CAPACITORS AND YOU SHOULD BE IN THE BALL PARK
TO BALANCE THE VOLTAGE.

THE PARTS NEEDED ALSO CAN BE FOUND AT WWW.
GRAINGER.COM AND THEIR PART NUMBERS ARE WHAT
WE HAVE LISTED BELOW. YOU CAN TAKE THESE
NUMBERS AND ANY SUPPLY HOUSE CAN CROSS
REFERENCE THEM FOR YOU TO THEIR PART NUMBERS.

K2 IS PART NUMBER 5X847, A STANDARD RELAY WITH
ONE SET OF "NORMALLY CLOSED POINTS" RATED AT
LEAST 30 AMPS AND HAS A 120 VOLT COIL.

K1 RELAY IS PART NUMBER 2A560 AND WE RECOMMEND
GETTING THIS FROM WWW.GRAINGER.COM . WE HAVE
HAD GOOD LUCK WITH THIS ONE.

CAPACITORS ARE DIFFERENT SIZES FOR DIFFERENT
HORSEPOWER OF MOTORS.

BUT A 4X662 PART NUMBER IS A GOOD ONE TO START
WITH. A GOOD RULE OF THUMB IS TO USE ONE 40-
50MFD CAPACITOR FOR EVERY ONE HP OF THE MOTOR.

SO, A THREE HP MOTOR WOULD NEED A TOTAL OF
ABOUT 120-150 MFD FOR THE STARTING CAPACITORS.

CAPACITORS HOOKED UP TO EACH OTHER IN PARALLEL
ADD TO EACH OTHER AND HOOKED IN SERIES THEY
SUBTRACT FROM THE TOTAL MFD.

IF YOU DECIDE TO TRY TO BALANCE THE OUTPUT
VOLTAGE OF YOUR MOTOR , YOU WILL USE WHATS
CALLED RUNNING CAPACITORS WHICH WILL BE
DISCUSSED LATER.   THE SAME RULE APPLIES HERE,
ABOUT A 40-50 MFD FOR EACH HP OF THE MOTOR.

IF YOU ARE RUNNING SENSITIVE EQUIPMENT, IE.,
(CMC MACHINE), YOU WILL PROBABLY NEED A POWER
TRANSFORMER TO ACCEPT THE VOLTAGE FROM THE
CONVERTER.

THIS TRANSFORMER WILL GIVE YOU THE OUTPUT
THAT YOU NEED TO RUN YOUR EQUIPMENT.

THE TRANSFORMER SIZE SHOULD BE LARGE ENOUGH TO
RUN THE TOTAL AMOUNT OF KVA THAT YOUR
EQUIPMENT NEEDS.IE.,2 HP. TO 35 HP. CONVERTER,
WILL NEED A 7 TO A 45 KVA. TRANSFORMER.

THE TRANSFORMER SHOULD BE A 3PH. TRANF. 440
VAC.Y TO 230 VAC. DELTA.

CHAPTER 7

HOW TO RUN ONE CONVERTER FOR THE ENTIRE SHOP:

THIS IS THE BEST WAY TO RUN THREE PHASE MOTORS ON SINGLE PHASE VOLTAGE.

WHAT YOU NEED TO DO FIRST IS TO DETERMINE THE LARGEST HORSEPOWER MOTOR YOU WANT TO RUN OFF OF THIS CONVERTER.

THEN THE ROTARY CONVERTER YOU WILL BUILD NEEDS TO BE LARGER THAN THIS MOTOR.

IF YOUR ROTARY CONVERTER IS A 10 HP MOTOR YOU CAN RUN A 7 1/2 HP MOTOR OFF OF IT OR A TOTAL OF TWO FIVES HP MOTORS, FOR EXAMPLE.

JUST DO NOT TRY TO START MORE THEN ONE AT A TIME.

START ONE AND THEN YOU CAN START THE NEXT ONE.

DO NOT OVERLOAD THE CONVERTER MOTOR.

IF THE MOTOR YOU WANT TO USE FOR A ROTARY CONVERTER HAS NINE LEADS, THEN USE THE DIAGRAMS ENCLOSED FOR THAT MOTOR AND BUILD THE CONVERTER AS SHOWN.

WHEN THIS IS DONE AND YOU HAVE THE CONVERTER MOTOR RUNNING, YOU USE THE LEADS OF "A", "B", "C" FOR THE THREE PHASE VOLTAGE TO SUPPLY TO OTHER MOTORS.

ITS A VERY SIMPLE SETUP.
AFTER THE CONVERTER IS RUNNING, YOU HAVE
ESSENTIALLY BUILT A THREE PHASE GENERATOR THAT
CAN BE USED TO START YOUR OTHER THREE PHASE
MOTORS AS LONG AS ITS LARGER THEN THEY ARE.

YOU SHOULD RUN THESE LEADS "A", "B", "C" TO
A 3 PHASE BREAKER PANEL AND THEN HAVE
INDIVIDUAL BREAKERS FOR EACH MOTOR YOU WANT TO
RUN.

IF YOU WANT TO RUN ANYTHING FROM A 120 VOLT
LINE OF THIS CONVERTER, SUCH AS A LIGHT BULB
OR WHATEVER. '

DO NOT TIE IT TO THE "C" LEAD. THAT VOLTAGE
VARIES AND WILL BURN UP WHATEVER YOU HOOK TO
IT.

YOU CAN ALSO BALANCE THE VOLTAGE OF "C" USING
A VOLTMETER AND ADDING WHAT IS CALLED RUNNING
CAPACITORS CONNECTED FROM LINE "C" TO LINE "A"
AND LINE "B".

START WITH THE SAME FORMULA AS BEFORE.

ABOUT 40-50MFD FOR EVERY HP OF THE MOTOR AND
WATCH THE VOLTAGE ON LINE "C" READING TO
GROUND.

WHEN ITS PRETTY CLOSE TO THE SAME VOLTAGE AS
"A" AND "B" YOU HAVE THE RIGHT SIZE RUNNING
CAPACITORS. ITS A TRAIL AND ERROR UNTIL YOU
GET IT CLOSE BECAUSE OF THE DIFFERENT
MANUFACTURES OF ELECTRIC MOTERS. BUT THIS IS
CLOSE.

The schematic below shows the electrical, as well as a mechanical floor plan, for one Rotary converter, running 2 three-phase electric motors.

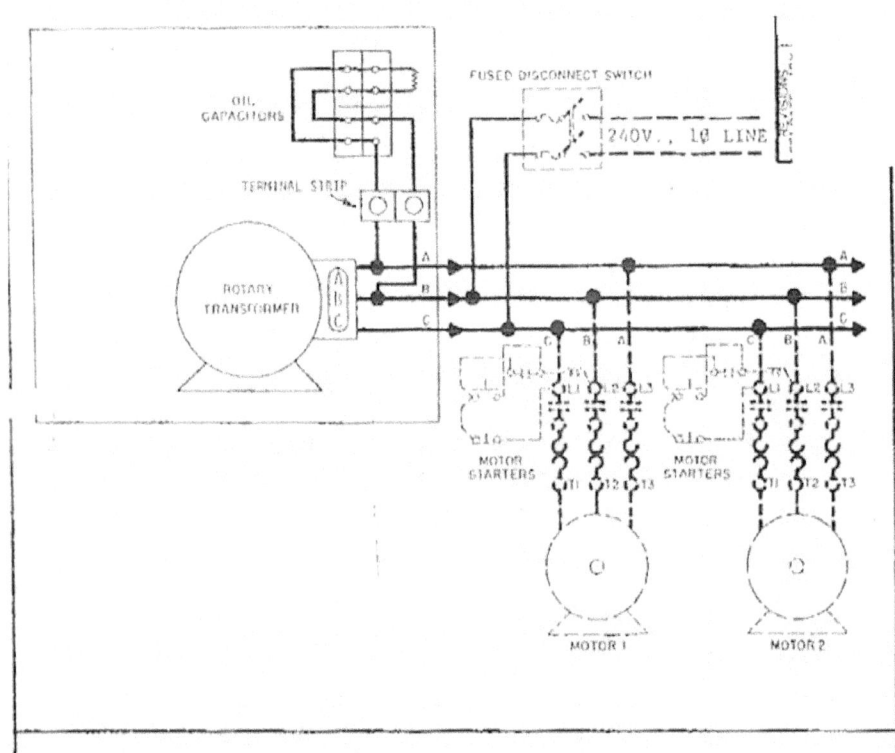

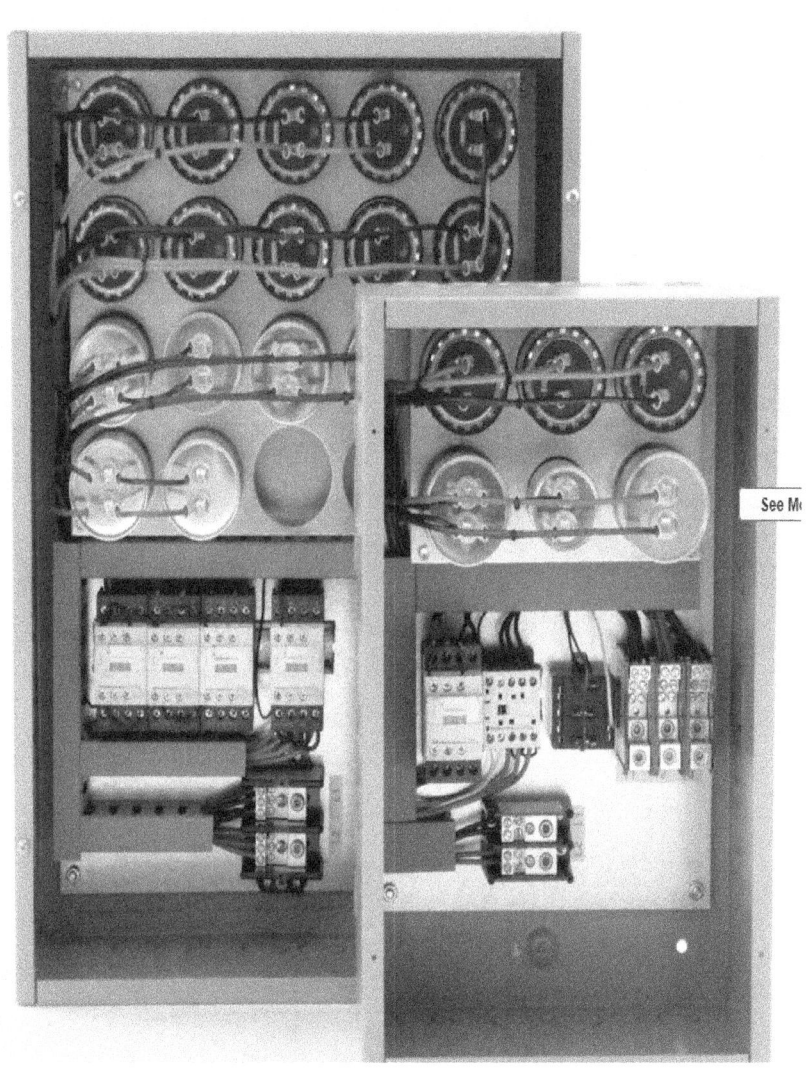

See M[...]

19

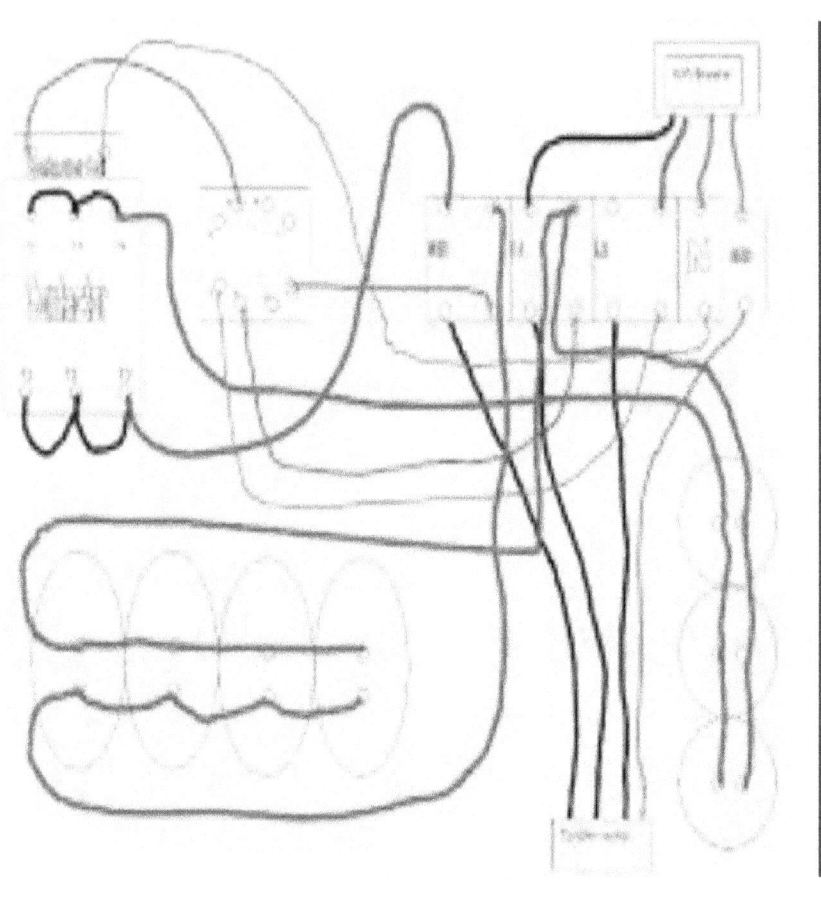

20

Chapter 8

IN CHAPTER 6-16 I MENTIONED THE FACT THAT THEY
MAY BE SOME APPLICATIONS, WHERE USING A PHASE
CONVERTER, YOU MAY NEED A POWER TRANSFORMER
FOR A STABLE OUTPUT ON ALL THREE LEGS. THE
PICTURE BELOW SHOWS WHAT ONE OF THESE
TRANSFORMERS LOOKS LIKE.

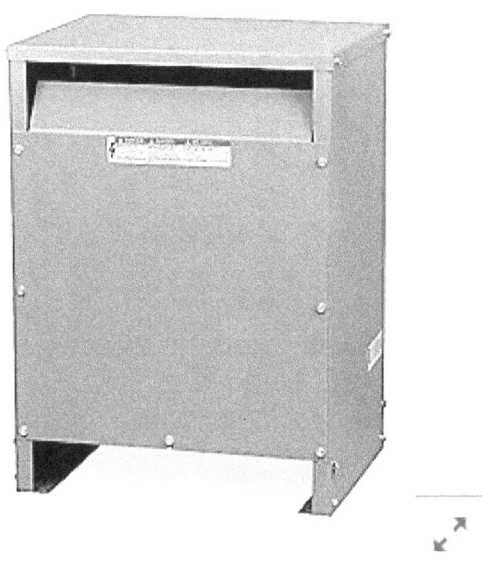

THE PLACES TO LOCATE THIS TRANSFORMER WOULD
BE, FIRST, TRY YOUR LOCAL ELECTRIC SUPPLY
COMPANY, WWW GRAINGER.COM AND OCCASIONALLY YOU
MAY FIND WHAT YOU NEED ON EBAY.

THE COMPANIES LISTED HERE ARE LOCATION IN
WHICH YOU MAY BUY THE PARTS, OR, YOU CAN
PURCHASE A UNIT COMPLETELY ASSEMBLED AND READY
FOR USE.

PHASE CONVERTER
GWM CORP. 800-437-4273

45KVA 3PH. TRANF. 440 VAC.Y TO 230 VAC. DELTA

TRANSFORMERS FOR PH. CONVERTERS

DOUG BEAT CO. 419-841-3881

DES ELECTRIC MOTORS & PHASE CONVERTERS
1722 TEXAS ST,   NATCHITOCHES, LA
318-352-8868   WWW.DESELECTRIC.COM
EMAIL DES@DESELECTRIC.COM

ALONG WITH THE PARTS SUPPLY LOCATIONS LISTED
ABOVE, YOU CAN FIND NUMEROUS OF PHASE
CONVERTER MANUFACTURERS ON GOOGLE.

BUT THE COMPANIES THAT I HAVE LISTED ABOVE, I
HAVE HAD GOOD LUCK DEALING WITH THEM.

www.ingramcontent.com/pod-product-compliance
Lightning Source LLC
Chambersburg PA
CBHW070737180526
45167CB00004B/1789